Panarchy

About Island Press

Island Press is the only nonprofit organization in the United States whose principal purpose is the publication of books on environmental issues and natural resource management. We provide solutions-oriented information to professionals, public officials, business and community leaders, and concerned citizens who are shaping responses to environmental problems.

In 2002, Island Press celebrates its eighteenth anniversary as the leading provider of timely and practical books that take a multidisciplinary approach to critical environmental concerns. Our growing list of titles reflects our commitment to bringing the best of an expanding body of literature to the environmental community throughout North America and the world.

Support for Island Press is provided by The Nathan Cummings Foundation, Geraldine R. Dodge Foundation, Doris Duke Charitable Foundation, Educational Foundation of America, The Charles Engelhard Foundation, The Ford Foundation, The George Gund Foundation, The Vira I. Heinz Endowment, The William and Flora Hewlett Foundation, Henry Luce Foundation, The John D. and Catherine T. MacArthur Foundation, The Andrew W. Mellon Foundation, The Moriah Fund, The Curtis and Edith Munson Foundation, National Fish and Wildlife Foundation, The New-Land Foundation, Oak Foundation, The Overbrook Foundation, The David and Lucile Packard Foundation, The Pew Charitable Trusts, The Rockefeller Foundation, The Winslow Foundation, and other generous donors.

The opinions expressed in this book are those of the author(s) and do not necessarily reflect the views of these foundations.

Panarchy
Understanding Transformations in Human and Natural Systems

Edited by Lance H. Gunderson and C. S. Holling

Synopsis by Bernice Wuethrich

ISLAND PRESS

Washington • Covelo • London

The development and production of this synopsis was underwritten by a grant from The Rockefeller Foundation.

This publication is a synopsis of the book *Panarchy: Understanding Transformations in Human and Natural Systems*, edited by Lance H. Gunderson and C. S. Holling. If you would like to order this book, please visit our Web site at www.islandpress.org.

6 x 9 inches, 507 pages
Tables, figures, index
Hardcover: $65.00 ISBN 1-55963-856-7
Paperback: $35.00 ISBN 1-55963-857-5

Manufactured in the United States of America
10 9 8 7 6 5 4

CONTENTS

PREFACE

Management of our natural resources, our forests and oceans, our lakes and rivers, and all their bounty often approaches its mission like a bullfighter entering a ring, ready to control and subdue nature. The theory of panarchy provides an alternative framework for managing the issues that emerge from the interaction between people and nature. That interaction generates countless surprises, the result of slow changes that can accumulate and unexpectedly flip an ecosystem or an economy into a qualitatively different state. That state may not only be biologically and economically impoverished, but also effectively irreversible.

Human and natural systems are complex, continually adapting through cycles of change. Change can be fast or slow—move with the speed of viruses multiplying or of mountains rising. It can take place on the scale of nanometers or kilometers. Change at one level can influence others, cascade down or up levels, reinvigorate or destroy. Yet most management zeros in on single scales and processes and quantifiable, focused goals. More often than not, transient success leads to larger failure.

Panarchy provides a framework to understand the cycles of change in complex systems, and to gauge if, when, and how they can be influenced.

The fundamentals of this new theory are developed in the book, *Panarchy: Understanding Transformations in Human and Natural Systems*, edited by Lance H. Gunderson and C.S. Holling. It reflects the work of the Resilience Alliance, a group of leading organizations and individuals involved in ecological and economic research around the world.

Panarchy is the result of an interdisciplinary effort of ecologists, economists, and social scientists. It emerged from expanding and integrating existing theories of complex adaptive systems, ecology, evolution, and organizational theory. Hundreds of scientists have participated, and a core group of about 25 contributed to the volume. "All the core members are brilliant in their own areas," Holling says. "Working together in the process of mutual cooperative discovery has added a wide breadth of vision."

Panarchy draws on Alliance members' experiences in multiple–use ecosystem management that span the globe, from the savannas of East Africa to the Great Lakes of North America, from the coral reefs of the South Pacific to the boreal forests of the sub-Arctic.

The following pages summarize the panarchy framework, drawing from the book, case studies, and interviews with leading ecologists and economists.

The work of the Resilience Alliance has been supported by a grant from the John D. and Catherine T. MacArthur Foundation to the University of Florida and the Beijer International Institute for Ecological Economics in Stockholm, Sweden. The Rockefeller Foundation has supported the application of the results described here in regional settings. The University of Florida also provided financial support for the project.

Panarchy

INTRODUCTION:
WHY PANARCHY?

There are few places on the earth more breathtaking or remote than the rainforest of Borneo. Its canopy is composed of high arcing dipterocarp trees that only reproduce during El Niño years. When the dry warmth of El Niño suffuses the region, dipterocarps across 60,000 square miles of forest reproduce in synchrony. They flower, fruit, and disperse their seeds within a 6-week period. The bounty creates a feeding frenzy on the forest floor, as orangutans, wild boar, parakeets, and other animals come to feast. Local people also gather the seed for home use or to sell. Yet the seeds are so abundant that some remain to sprout.

Now that unique system is breaking down. Deep within a protected national park on Indonesian Borneo, many dipterocarp species have failed to produce a single seedling since 1991. Researchers blame interactions between linked natural and human systems: El Niño-induced drought and deforestation.[1]

In the last two decades, the El Niño cycle has changed, becoming longer and more severe, a trend scientists think is linked to global climate change. The forest trees are not adapted to such prolonged, severe droughts, and the quantity and quality of their seed has fallen. In addition, unprotected forests are heavily logged, denuded of trees, or destroyed by fire. Therefore, wildlife in these areas stream into the ever-smaller tracts of intact forest when seeds fall. They devour all the seed, leaving none to regenerate—and still animals go hungry. Impaired seed production can no longer keep up with wildlife, and animal populations, including the endangered orangutan, are plummeting.

The human and natural worlds are inextricably linked. The tighter the coupling, the greater the possibility of sweeping changes. In Borneo, the potential collapse of an entire ecosystem and its inhabitants is coupled to global climate change, and to a global appetite for lumber that is deforesting the landscape. In turn, the loss of the forest could have a huge financial impact, since timber exports contribute as much as $8 billion annually to the Indonesian economy. This exported wood also amounts to 80 percent of plywood used in the U.S. home construction industry, which would have to look elsewhere for raw materials should the ecosystem flip into a degraded state. Housing costs in the U.S. could rise; the ecological pressures of wood production could shift from one region to another.

Indeed, cascades of change are reshaping natural and human systems around the globe. Novel diseases have emerged in socially and ecologically disturbed areas of the world. In the case of HIV, the increased mobility of people and a lack of human foresight have transformed the virus into a global epidemic. The collapse of the Soviet Union still reverberates politically, economically, and environmentally as new nations and national alliances splinter and coalesce. The Internet is altering the conduct of commerce, science, and culture. The rise of supranational economic organizations that shape development, trade, and the environment has provoked a deeper popular involvement in global concerns.

Change is coming faster and faster. Grasslands are becoming deserts; rainforests are being turned to pasture; coral reefs are dying. In the 1980s, coral reefs in the Caribbean collapsed, declining to less than 2% cover from 50% or more.[2] Approximately half of the world's wetlands have been lost in just the last century. Topsoil needed for farming is disappearing. In the last 300 years, the rate of topsoil loss was 300 million tons per year; in the last 50 years it has more than doubled, to 760 million tons per year.[3]

"The scale at which humans impact the planet is unprecedented," says Resilience Alliance member Steve Carpenter, an ecologist at the University of Wisconsin. "We're using about half the primary production in the world and about half the fresh water. We've used more than half of the tillable land and more than half of marine fisheries. Change is accelerating and we have the potential for even more abrupt change in the coming years."

These trends augur great uncertainty and unpredictability. They nonetheless allow the possibility that human foresight and innovation can develop paths that sustain natural diversity and improve human well-being. Panarchy is an integrative theory to help understand the source and role of change in systems, and to identify development paths that are truly sustainable.

For example, Resilience Alliance ecologists Nick Abel, Brian Walker, and other colleagues in western New South Wales, Australia, have worked together with aboriginal peoples, pastoralists, conservationists, and the minerals and tourism industries to develop visions of sustainable land use patterns in rangelands that have been heavily degraded since European settlement. "We are using the ideas of panarchy to analyze the rangelands, and understand what and why undesirable changes are taking place," Walker says. Policymakers from 45 organizations participated in workshops that designed institutional changes needed to support sustainable development.

But why is panarchy likely to succeed in creating sustainable futures when the road is strewn with so many failures? As a tool to affect change in linked ecological and social systems, panarchy differs in fundamental ways from dominant management approaches:

- Panarchy is integrative. Depending on the problem at hand, it incorporates a range of disciplines from the biophysical and social sciences. It eschews the artificial separation between disciplines that has dominated the development of science and society, as well as the separation of humans and nature.

- Panarchy approaches ecological and social systems as complex adaptive systems in which components adapt to and change with their environments, leading to unpredictable outcomes. It allows management with the expectation of surprise. In contrast, dominant management approaches consider ecosystems as linear, predictable systems capable of maintaining an ideal state of equilibrium.

- Panarchy considers system function as a whole and aims to maximize the system's resilience. In contrast, prevailing "command and control" approaches focus on controlling a single variable, such as fish yield. These approaches manage ecosystems to conform to human requirements, rather than learning how humans can adapt as part of ecological systems.

- Panarchy emphasizes the importance of relevant interactions across geographic and time scales. Most management approaches focus on a single scale.

- Panarchy simplifies complexity. "Behind the great complexity of socio-economic processes beats a heart of simple operation," says C.S. Holling, an ecologist at the University of Florida. That essentially simple structure can be modeled recognizing just three to five key variables, and the significance of the speeds and spatial scales at which they operate.

To describe the cross-scale and dynamic character of interactions between humans and nature, Holling and co-editor Lance Gunderson of Emory University coined the term *panarchy*. It draws on the Greek god Pan, a symbol of universal nature, to capture an image of unpredictable change. It draws on the notion of hierarchies across scales to represent structures in natural and human systems that sustain system integrity, allow adaptive evolution, and at times succumb to gales of change. Hence the word panarchy.

Panarchy grounds itself in an appreciation of uncertainty and the expectation of surprise. It flows from a respect for resilience and learning, for memory and adaptation, and for the novelty that emerges to trigger renewal. Fundamental to panarchy is an understanding of the adaptive cycle—a pattern of rapid, opportunistic growth, conservation, destruction, and renewal—and the concept of scale. Furthermore, panarchy calls for innovative approaches to management, called adaptive management, based on learning by doing. The following sections will explore these topics.

CHAPTER 1
THE ADAPTIVE CYCLE: SURPRISE AND RENEWAL

Life proceeds through uneven rhythms of change—slow periods of gradual change and sudden surprises. The surprises can be negative or positive: an industry is made suddenly obsolete by the development of a new technology; an insect pest erupts in a forest; a hobby transforms into a thriving business; a series of genetic mutations allow an animal to fight off a parasitic infection.

"Such surprises are an essential part of any living dynamic system," says Holling. "Through surprises, systems are both renewed and tested. Within ecosystems, biosurprises play the enormously important role of introducing unexpected novelty, a potential source of renewal."

Understanding when and how novelty emerges or is suppressed lies at the heart of panarchy. It can provide policymakers and ecosystem managers with vital insights into when and how to act—and whether action is fruitless.

"Management of natural resources goes in cycles," Gunderson says. "At times in the cycle there is leverage to change things. What people usually see in management systems are the gridlock and the failure. The panarchy cycle points out that there are phases marked by opportunity, creativity, and novelty—when good things can happen."

By studying dozens of ecosystems around the world over the last several decades, ecologists have learned that novelty emerges as part of a cycle consisting of four phases: *rapid growth, conservation, release* (or "creative destruction"), and *renewal*. From one phase to the next, the strength of a system's internal connections—its flexibility, resilience, and its vulnerability to disturbance—change. The adaptive cycle is not an absolute, and many variations exist in human and natural systems. However, it provides both a useful metaphor to classify systems and order events and a theoretical framework in which to pose questions and testable hypotheses relevant for understanding transformations in linked systems of people and nature.

Although ecologists have most thoroughly documented the adaptive cycle, the idea was sparked by an Austrian economist, Joseph Schumpeter, whose writings span the first half of the twentieth century. Analyzing the economy's boom and bust cycles, he described capitalism as a "perennial gale of creative destruction," coining the phrase now used to describe the disturbances that periodically punctuate the adaptive cycle. A closer look at the

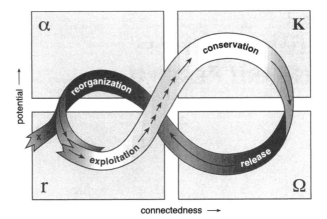

Figure 1. A stylized representation of the four ecosystem functions (r, K, Ω, α) and the flow of events among them. The arrows show the speed of that flow in the cycle, where short, closely spaced arrows indicate a slowly changing situation and long arrows indicate a rapidly changing situation. The cycle reflects changes in two properties: (1) Y axis—the potential that is inherent in the accumulated resources of biomass and nutrients; and (2) X axis—the degree of connectedness among controlling variables. Low connectedness is associated with diffuse elements loosely connected to each other whose behavior is dominated by outward relations and affected by outside variability. High connectedness is associated with aggregated elements whose behavior is dominated by inward relations among elements of the aggregates, relations that control or mediate the influence of external variability. The exit from the cycle indicated at the left of the figure suggests, in a stylized way, the stage where the potential can leak away and where a flip into a less productive and organized system is most likely.

adaptive cycle sheds light on how it may operate in both ecosystems and economic or social systems.

Four Phases of the Adaptive Cycle

1. *The rapid growth or r phase.* Early in the cycle, the system is engaged in a period of rapid growth, as species or other actors colonize recently disturbed areas. These species (referred to as r-strategists in ecosystems), utilize disorganized resources to exploit every possible ecological niche. The system's components are weakly interconnected and its internal state is weakly regulated.

   In ecosystems, the most successful r-strategists are able to proliferate despite environmental variation and tend to operate across small geographical areas and over short time scales. In economic systems, r-strategists are the innovators and entrepreneurs who seize upon opportunity. They are start-ups and producers of new products; they capture shares in newly opened markets and initiate intense commercial activity.

2. *The conservation or K phase.* Transition to the K phase proceeds incrementally. During this phase, energy and materials slowly accumulate. Connections between the actors increase. The competitive edge shifts from species that adapt well to external variability and uncertainty to those that reduce its impact through their own mutually reinforcing relationships. These "K-strategists" operate across larger spatial scales and over longer time periods. As the system's components gradually become more strongly interconnected, its internal state becomes more strongly regulated. New entrants are edged out while capital and potential grows, and the future seems ever more certain and determined.

   In an ecosystem, the potential that accumulates is stored in resources such as nutrients and biomass. An economic system's potential can take the form of managerial and marketing skills, accumulated knowledge, and inventions.

   But the growth rate slows as connectedness increases to the point of rigidity and resilience declines. The cost of efficiency is a loss in flexibility. Increasing dependence on existing structures and processes renders the system vulnerable to any disturbance that can release its tightly knit capital. Such a system is increasingly stable, but over a decreasing range of conditions. The transition from the conservation to the release phase can happen in a heartbeat.

3. *The release or omega (Ω) phase.* A disturbance that exceeds the system's resilience breaks apart its web of reinforcing interactions. In an abrupt turnabout, the material and energy accumulated during the conservation phase is released. Resources that were tightly bound are transformed or destroyed as connections break and regulatory controls weaken. The destruction continues until the disturbance exhausts itself.

   The disturbance can occur when a slow variable triggers a fast variable response. For instance, the slow growth and aging of a fir forest triggers the outbreak of an insect pest. Or the slow growth of debt and gradual decline of profits finally triggers a financial panic.

   In ecosystems, agents such as forest fires, drought, insect pests, and disease cause the release of accumulations of biomass and nutrients. In the economy, a new technology can derail an entrenched industry. But the destruction that ensues has a creative element. This was Schumpeter's "creative destruction." Tightly bound capital—whether equipment, money, skills, or knowledge—is released and becomes a potential source of renewal.

4. *The renewal or alpha (α) phase.* Following a disturbance, uncertainty rules. Feeble internal controls allow a system to easily lose or gain resources, but it also allows novelty to appear. Small, chance events have the opportunity to powerfully shape the future. Invention, experimentation, and re-assortment are the rule.

In ecosystems, pioneer species may appear from previously suppressed vegetation; seeds germinate; non-native plants can invade and dominate the system. Novel combinations of species can generate new possibilities that are tested later.

In an economic or social system, powerful new groups may appear and seize control of an organization. A handful of entrepreneurs can meet and turn a novel idea into action. Skills, experience, and expertise lost by individual firms may coalesce around new opportunities. Novelty arises in the form of new inventions, creative ideas, and people.

Early in the renewal phase, the future is up for grabs. This phase of the cycle may lead to a simple repetition of the previous cycle, or the initiation of a novel new pattern of accumulation, or the precipitation of a collapse into a degraded state.

Taken as a whole, the adaptive cycle has two opposing stages. The "front loop" encompasses rapid growth and conservation, and the "back loop" encompasses release and reorganization. The front loop is characterized by the slow accumulation of capital and potential, by stability and conservation. The back loop is characterized by uncertainty, novelty, and experimentation. The back loop, and the renewal phase in particular, is the time of greatest potential for the initiation of either destructive or creative change in the system. It is the time when human actions—intentional and thoughtful or spontaneous and reckless—can have the biggest impact.

An Industrial Cycle

The adaptive cycle metaphor may shed light on aspects of the auto industry's history, says Resilience Alliance member William Brock, an economist at the University of Wisconsin in Madison. In the 1950s, when gas was very cheap and gas wars kept driving its price even lower, Detroit automakers were building large tailfinned cars. They were gas-guzzling machines, glamorous and big, Brock recalls. The automobile industry at that time could be considered in the conservation phase of the adaptive cycle, accumulating capital, building factories, and growing ever more dependent on the production of large automobiles, until it was set on cruise control and the future looked great.

Then, in the early 1970s, gas prices skyrocketed, and cars were lining up for blocks to get gasoline. People suddenly became interested in buying small cars made in Japan. The unexpected foreign competition sent the automobile industry into a tailspin, abruptly ushering in the release phase. The industry's first response was to try to convince the U.S. government to impose high tariffs and quotas on Japanese imports. This step would inflate the price of foreign-made cars and discourage American buyers, thereby attempting to protect an American industry that had been caught asleep at the wheel. "But the demand in the U.S. for smaller cars not only pressured the political system to get out of the way, it also pressured the U.S. industry to reevaluate its strategy," Brock says. Using the adaptive cycle metaphor, this reevaluation could be viewed as a renewal phase, when novelty arises and new ways of doing business and meeting demand can emerge.

Indeed, GM, Ford and Chrysler regeared. Corporate executives, labor unions, and creative teams were now open to new ideas. They designed new cars, reorganized their assembly lines, and reformed relationships between labor and the companies. Old "capital" in the form of old ways of doing business disappeared in a gale of creative destruction. Gone were the tailfins of old; light, fuel-efficient, lightweight sedans ruled the day. U.S. automobiles rose in quality and began to meet the public's needs, ushering in a new cycle—which appears to be on a course similar to the last.

Cars have gotten bigger again, SUVs rule, and gas prices are once more rising, casting doubt on the maxim that bigger is better. "The adaptive cycle metaphor suggests the industry is heading toward another round of creative destruction," Brock says. He predicts that within a few years, we may have more expensive gas or gas pump lines. The auto industry may have to pass through another crisis and before it starts building new products like hybrid, smaller SUVs that can run on 40 miles per gallon.

An Ecosystem/Management Cycle

A second example of the adaptive cycle comes from a classic ecological study of spruce/fir forests that grow across a huge swath of North America, from Manitoba to Nova Scotia and into northern New England. Among the forests' many inhabitants is the spruce budworm, a moth whose larvae eat the new green needles on coniferous trees. Every 40 to 120 years, populations of spruce budworm explode, killing off up to 80% of the balsam firs. But renewal follows this episode of creative destruction, and the forests regrow to repeat the cycle.

Following World War II, a campaign to control spruce budworm became one of the first huge efforts to regulate a natural resource using pesticide spraying. The effort, supported by an air force in New Brunswick, was meant to minimize the economic consequences of the pest on the forest industry. It ran into problems early on.

"The goal was not to eliminate the insect but to keep the forest green, which, unfortunately, is good for budworm too, since they like mature trees," Holling explains. While the moderate spraying regime avoided a catastrophic budworm outbreak, it allowed the insect to grow in numbers and range, until it became a simmering problem spread across ever-larger areas. Meanwhile, the program's partial success increased industry's dependence on the spraying program, intensified logging, and spawned more pulp mills. "Management was trapped—if they stopped spraying there would be an extensive impact. It was the standard resource management pathology," Holling says.

In 1973, Holling, working with the International Institute of Applied Systems Analysis in Vienna, tackled the problem. A series of workshops, and the leadership of Gordon Baskerville, the newly elected minister of the Department of Natural Resources, enabled a reinvention of policy.

The advances were made based on an understanding of the natural adaptive cycle of forest and pest. The cycle begins in the rapid growth phase, when the forest is young. Trees are small and the foliage that budworm larvae consume is limited. While spruce budworm is present, forest birds that prey on the larvae keep its numbers down. During the transition from rapid growth to conservation, the forest slowly matures and foliage increases, providing more food for the budworm. The bird population increases as well, but tops out while the forest canopy continues to grow. Finally, during the cycle's release phase, the larvae outstrip the ability of the birds to control them. Larvae numbers explode, killing the majority of forest trees. Their rapid demise opens up new opportunities for plants to grow, and during the renewal phase, the forest ecosystem begins to reestablish itself. The cycle repeats.

Understanding this cycle led to fundamental changes in management of the forest and the pesticide regimen. Rather than using low doses, the foresters needed big doses, but only when the budworm population had exploded beyond the capacity of natural controls to keep their numbers in check. Along with an appropriate harvesting regime, these less-frequent doses of pesticides represented a way to work with nature at both the beginning of the cycle and at its end. The result was that the forestry industry introduced much less insecticide into the ecosystem overall. The insights also changed the way in which the industry assessed the status of the forest and the way the forest was managed by different companies. In essence, the industry gained more freedom locally to develop innovative ways to harvest mature trees to better compete with the budworm. Management of the forest was thus fundamentally transformed, regional leadership combined with local independence to achieve more ecologically responsible harvesting of the forest.

The example of the spruce budworm and the fir forest illustrates a natural disturbance to which an ecosystem has adapted. Companies that harvested the forest initially operated without regard to the natural cycles of trees and pest. Early policy was headed for disaster. But management went through its own adaptive cycles and was able to learn and change, not only averting disaster, but developing sustainable forestry practices in harmony with natural cycles. Unfortunately, such successes are more the exception than the rule.

CHAPTER 2

THE PATHOLOGY OF RESOURCE MANAGEMENT

Like the spruce/fir forest, other ecosystems are adapted to disturbances ranging from fire to storms, and usually recover rapidly. But now human-induced disturbances commonly occur on a far greater scale than natural disturbances. Increasingly, these human disturbances—including the effects of management—exceed the resilience of the system.

Most ecosystem management focuses on maximizing the output of a particular product, such as fish catch or wheat yield. By managing with the goal of increasing the productivity of a narrowly defined product, the overall resilience of the system is weakened, making it rigid and more vulnerable to disturbance. It becomes an accident waiting to happen. At the same time, management agencies typically become more myopic and rigid, relevant industries become more dependent and inflexible, and the public trust is lost.[4] Overly stressed ecosystems lose their integrity and flip into a degraded state. That state, however, may be as stable as its preferred counterpart, and may persist for decades or even centuries.

Gunderson and Holling call this now familiar sequence of events the "pathology of resource management." Consider how:

- Semi-arid rangelands become shrub deserts: moderate, stable grazing by cattle reduces the diversity of rangeland grasses. Among the grasses lost are drought-resistant species. Their disappearance means that droughts leave the soil denuded of plants, causing the soil to become less permeable and reducing the amount of available water.

- Flood control and irrigation lead to large ecological and economic costs: effective flood control leads to higher human settlement in fertile valleys and a large investment in vulnerable infrastructure. When a large flood eventually overwhelms the dam and dikes, the result is a dramatic reconfiguration of the river's social and economic landscape.

- Fisheries collapse in spite of a highly developed theory of fisheries management. The North Atlantic cod fishery is a typical example. Its initial success led to increased investment in fishing fleets and equipment and over-exploitation of the fish. When the fish stock showed signs of distress, management agencies became paralyzed. The collapse of the fishery in the early 1990s put 30,000 Canadians out of work and ruined the economies of 700 communities.

In each case, management targets a single variable (meat production, water levels, and fish stock, respectively) to control at optimal levels. It presumes to have replaced uncertainty in nature with the certainty of human control. Social systems initially flourish from this ecological stabilization and resulting economic opportunity. But that success creates its own failure.

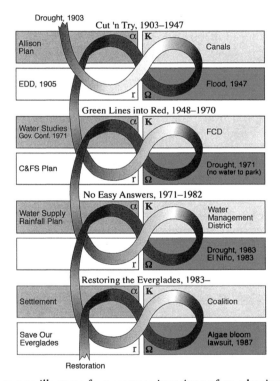

Figure 2. Fours eras illustrate four separate iterations of an adaptive cycle of water management in the Florida Everglades. Each management era is characterized by a slow period of capital accumulation, followed by a perceived crisis and reformation. The first of these eras began in 1903 with efforts to dig canals to drain the system for development agriculture in a strategy labeled "Cut 'n Try." The second and most prominent era (1948–1970) involved massive federal and state public works projects to prevent flood damage, and was dubbed "Turning Green Line to Red." The third era, "No Easy Answers" (1971–1982), was an attempt to restructure existing management agencies into a new system-wide agency to deal with both water shortage and flood control. The most recent era, beginning in 1983, is characterized by attempts to restore the natural values of the system.

Stabilization of the target variable leads to slow changes in other ecological, social, and cultural components—changes that can ultimately lead to the collapse of the ecosystem, the management system, and larger components of the economic and social systems as well.

The history of the Everglades is a prime example of the pathology of management. Over the course of the last century, policy and organizational changes in the Everglades' management has occurred in spasmodic lurches of learning driven by crises. The crises themselves were precipitated by earlier myopic policies, whose initial successes led to larger failures.

"Rather than take advantage of nature's dynamics, we humans have repeatedly tried to impart control, at an ever-mounting effort and cost," says Gunderson. He, Holling, and Garry Peterson at the University of Wisconsin analyzed that management history[5] as an example of a strongly linked ecological and social cycle.

Cycles of Crisis and Learning in the Everglades

The Everglades, often described as a "river of grass," has been dredged, drained, and dammed. In a vain attempt to control its flow, four successive cycles of management constructed 720 miles of levees, 1,000 miles of canals, 16 pumping stations, and 200 gates to control water. Each of the four cycles experienced periods of slow growth and policy implementation interrupted by ecological crises that led to new configurations of water management institutions. Each new configuration can be thought of as an alternative stable state of the management system.

Historically the Everglades comprised 4,000 square miles of subtropical wetland, from Lake Okeechobee in the north to the Florida Bay at the south. It is a patchwork of landscapes: lakes and rivers, freshwater marshes, tree islands, mangrove swamps, pinelands, and coastal waters. Through it all, water flows. It begins in the northern end at the headwaters of the Kissimmee River, fed almost entirely by rain. Rain drenches the system every summer with an average of 40 to 65 inches of water. This water flows into Lake Okeechobee, which is wide and shallow, covering close to 730 square miles and averaging just 12 feet in depth.

Rainy season deluges historically flooded the lake's southern shore, forming the "river of grass" as flowing waters up to 50 miles wide and 100 miles long inundated the landscape on their way to the ocean. The species of the Everglades evolved in concert with an environment that was nutrient-poor and episodically flooded. Many of these are unique to the system.

The First Cycle

In the early 1900s, the Everglades was viewed as a big mean swamp. As author James Carlos Blake describes it, "If the Devil ever raised a garden the Everglades was it."[6] That view was seemingly borne out by the flood of 1903, which destroyed most of the region's farms. In 1906, Florida governor

Napoleon Bonaparte Broward began to make good on a campaign promise to "Drain that abominable, pestilence-ridden swamp."[7] He initiated the first engineering campaigns to reroute the wetlands' waters to the Atlantic and dry out the land for farming. The overseeing institution was the Flood Control District, in which a board of trustees managed the building of canals and levees.

But hurricanes from 1926 to 1928 brought more flooding and more death.[8] The Army Corps of Engineers responded with a major flood control project, constructing levees that bounded Lake Okeechobee like a Berlin Wall. Reassured by brick and mortar, agriculture reestablished itself. During these years, the management system cycled through rapid growth and conservation. Increasing structure in the form of canals reflected an attempt to control the natural system. The release phase was unleashed by tropical storms and hurricanes that hit South Florida in 1947 and 1948. More than 2,000 people and 25,000 cows drowned. An incredible 108 inches of rain fell on the region in six months, submerging farm, field, and town.

The Second Cycle

The crisis gave birth to new era of management control. Under a large bureaucratic plan called the Central and Southern Florida Project for Flood Control and Other Purposes (C&SF), the Corps erected levees, canals, and pumps capable of controlling 13.8 billion liters of water per day. The C&SF put in place operational procedures and created dedicated land-use areas: the Everglades Agricultural Area, Water Control Areas, and the Everglades National Park. It also sparked a surge in South Florida's human population. The C&SF reflected a partnership of federal, state, and local governments in control of the hydrologic infrastructure. It became more and more interconnected, building with a rigidity that primed the Everglades for disaster.

Surprise came in the 1960s and 1970s with the arrival of droughts—not the planned-for floods. The dried-out watershed erupted in flames. Through the heat and smoke came the realization that decades of management had mismanaged the Everglades and were destroying the natural system.

The Third Cycle

This began in 1971, with attempts to put in place mechanisms to deal with water shortages as well as floods. The emphasis during this cycle was on creating social rather than physical capital through an institutional reconfiguration that formed the South Florida Water Management District.[9] But droughts and flooding continued in the early 1980s, until the problems were so severe—and the public outcry so great—that Governor Bob Graham initiated a "Save Our Everglades" campaign. This marked the onset of the fourth cycle.

The Fourth Cycle

Driven by fears of pollution, as Lake Okeechobee suffocated under a huge algae bloom, the management structure again metamorphosed, and gave rise to the Everglades Coalition. This alliance of governmental and nongovernmental organizations aimed to resolve chronic environmental issues such as the agricultural runoff that was choking the lake. This cycle concluded with the settlement of a water quality lawsuit in 1991.

By the time the gears were in motion for serious reevaluation of policy, successive eras of management had reduced the Everglades in size by 50%, reduced flows to the existing Everglades by 70%, and grossly undermined water quality. Today, 68 species in the Everglades are endangered, including the Florida panther, the manatee, the snail kite, and the Southern bald eagle. Agriculture and the human population continue to grow, and along with them, the need for water.

The current era began in 1996 with a study that led to passage of the Comprehensive Everglades Restoration Plan, or CERP, in 2000. It includes dozens of projects estimated to take 36 years to complete, at a price tag of $8 billion. Unlike its predecessors, today's era is based on a sound ecological understanding of the system. It marks the first time the U.S. government has reversed a major public works program out of consideration for the environment. Yet, while success is possible, it is far from assured, as a command and control mentality still persists, Gunderson cautions.

On the other hand, the Everglades' cycles also reflect the ability of humans to learn and to rapidly create novelty in the form of system-wide responses. "Social systems can invent new ways of dealing with changes in the environment. The diversity of adaptations may ultimately succeed in dynamically managing ecosystems as complex as the Everglades," Gunderson says.

CHAPTER 3
RESILIENCE

People use the term resilience to mean many things, usually the ability of something to bounce back to its previous condition. In panarchy, however, resilience means something subtly different. It refers to the ability of a biological system, an ecosystem, or a social system to withstand disturbance and still continue to function. A measure of resilience is the magnitude of disturbance that can be experienced without the system flipping into another state.[10]

Consider the human heart, designed to last a lifetime. With each beat it pumps oxygenated blood to the whole body, including the heart itself. Many factors help the heart in its work and support its natural resilience. The maintenance of healthy blood pressure keeps the heart steady. Blood pressure can be thought of as the weight that the heart needs to lift—if blood pressure is low, the weight is light, and the normal heart has an easy job.

In a heart attack, an artery supplying the heart is blocked, and part of the heart is deprived of its own life's blood. A moderate heart attack may kill part of the heart muscle, but leave enough healthy tissue for the heart to beat. A series of heart attacks will progressively weaken the heart, eroding its resilience—its ability to withstand further insult. Blood pressure falls as the heart pumps less strongly. But this fall does not lessen the load on the heart. On the contrary, below a certain threshold, it triggers a maladaptive response in the kidneys, which mistakenly interpret the low flow of blood as dehydration. The hormonal cascade that ensues causes arteries to clamp down and makes the already weakened heart work harder still. The added strain exacerbates the failure of the heart, whose natural resilience has finally been overcome.

Although stabilizing drugs may be administered, the heart will never regain its normal strength. It operates in an altered, qualitatively weaker state. It needs new chemical inputs to function at all, and reorganizes its own internal controls based on dependence on the medications that maintain the interconnected function of the heart, the arteries, and the kidneys. Depending on the extent of the damage, the heart may continue weakly pumping for hours or years. But there is no going back.

Panarchy recognizes that once a certain threshold is passed, it is not possible to bounce back. In an ecosystem, recovery may be impossible at worst,

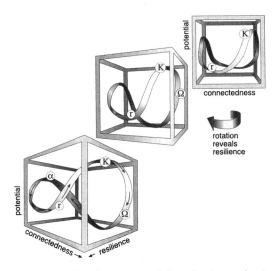

Figure 3. Resilience is another dimension of the adaptive cycle. A third dimension, resilience, is added to the two-dimensional box of Figure 1, showing that resilience expands and contracts throughout the cycle. Resilience shrinks as the cycle moves toward K, where the system becomes more brittle. It expands as the cycle shifts rapidly into a "back loop" to reorganize accumulated resources for a new initiation of the cycle. The appearance of a figure 8 in Figure 1 is shown to be the consequence of viewing a three-dimensional object on a two-dimensional plane.

an expensive and distant hope at best. The new, degraded state may persist for decades or centuries. The preferred and disturbed states are known as alternative stable states. As far as nature is concerned, both are stable and able to carry on. Panarchy approaches ecosystem management by identifying the preferred state, and seeking ways to enhance its resilience.

"People tend to want to stabilize nature for economic purposes. By doing that, we tend to lower the resilience of systems—which means that they constantly surprise us by changing state," Gunderson says.

Panarchy asks how resilience can be lost, what can be done to enhance it, and whether there are equivalent properties in social and economic systems.

CHAPTER 4

CONNECTEDNESS

The strength of connections between elements within a system moderates the influences of the outside world. An organism, ecosystem, organization, or economic sector with high connectedness can ride the waves of external variability; internal regulatory processes largely control its operation and fate.

A biological example of strong connectedness is temperature regulation in endothermic or "warm-blooded" animals. Five physiological mechanisms (such as evaporative cooling and metabolic heat generation) keep the organism's internal temperature within a narrow range, regardless of how hot or cold it is outside. Each regulatory mechanism operates over a specific range of temperature with different efficiencies and feedback controls. The result is a remarkably robust regulation of body temperature. But a large enough disturbance—whether generated internally (for example, from alcohol consumption) or externally (from frigid air)—can overwhelm the regulatory mechanisms or break down the internal connections.

In ecosystems, the connectedness between a mix of biological and physical interactions forms mutually reinforcing relationships. These relationships give rise to sustaining structures and processes that reinforce their own expansion and functionality. Such a system is called "self-organized."

An example is found in woodlands that give hillsides a tiger-stripe pattern in semi-arid parts of Australia, Africa, and Mexico. Called mulga woodlands in Australia, the stripes move across the landscape over the course of decades. They form and slowly move as a result of the interactions between water, soil, vegetation, and topography. As water flows down a hillside it encounters vegetation. The plants slow the water down and it soaks into the ground, where it becomes available to plant roots and encourages further plant growth. Plants that prosper along the uphill, well-watered edge impede the further flow of water. Little water is then available to plants at the downhill edge, which suffer increased stress. Beyond this far edge, the water is exhausted and the soil is bare. Alternating strips of green vegetation and bare land form the striped pattern. It slowly migrates uphill because at the moister, uphill side of the stripe, bare areas are colonized, while along the downhill side vegetation dies off.[11] If trees are cut and their water-concentrating effects are lost, the entire system becomes less productive, because in the uniform system that results there is not enough water available at any point for much to grow.

On the one hand, an organism, ecosystem, or organization with high connectedness is less influenced by external variability. On the other hand, highly self-organized systems may be more sensitive to disturbance than are loosely connected systems. If the web of interconnections in a self-organized system is weakened at a crucial point, it may unravel. Therefore, as connectedness grows, so, too, does rigidity.

CHAPTER 5
MATTERS OF SCALE

Pathological management led to a loss of resilience in the Everglades, and part of the problem was management's narrow focus on processes at a single scale, whether in space or time. According to Gunderson, most schemes to manage water in the Everglades have focused on the annual cycle of rainfall, in which 85% of the rain falls in the three brief months of summer. Legislation passed in 1970, for example, was intended to guarantee a minimum annual water delivery to Everglades National Park. That minimum was 350,000 acre-feet per year, and every year over the next decade, the park received the legislated amount, regardless of how much rain actually fell.

The policy ignored longer multi-year cycles of rainfall that are linked to El Niño. "Rather than think about managing the system in multiple scales, much of resource management is fixed on short-term cycles," Gunderson says.

Cross-scale influences are everywhere. Global climate change affects regional ecosystems and local human health and livelihoods. New diseases like AIDS can emerge and rapidly cross scales, speeded along by forest clearing, the movement of refugees, and international travel. Economic globalization affects the regional environment and local employment.

Forecasting the weather; taking a candidate vaccine from the laboratory bench to mass production; predicting carbon dioxide flux across a landscape based on experiments inside a plastic bubble—all involve questions of scale. Much of science assumes that processes are scalable, and that predictions can be made into the future and across spatial scales. Panarchy argues that the limits to scalability are a major source of unpredictability in ecosystem management.

"To strengthen ecosystem resilience, we need to integrate processes that cover meters to tens of thousands of kilometers, minutes to millennia," Holling says. "Management has to be flexible, adaptive, and experimental at scales compatible with those of critical ecosystem functions and social processes."

Linked scales form a type of hierarchy that is unlike standard notions of top-down ecological hierarchies in which large, slow processes control everything below them. In nature, hierarchical levels are semi-autonomous, dynamic structures in which smaller, faster processes nest inside and interact with larger, slower ones. Between scales, there is typically at least an order of magnitude's difference in speed and in size. Each scale of the hierarchy has its own adaptive cycle, which interacts with the levels above and below. These nested structures and processes form panarchies.

One example of cross-scale effects discussed earlier is found in the rain-forests of Borneo, where global climate change appears to be driving regional climate change. In addition, global demands for lumber coupled with local poverty are leading to extensive logging and clearing of land. Forest ecologist Lisa Curran at the University of Michigan has shown that the combination of these factors is derailing the reproductive cycle of this unique ecosystem.

In addition to climate and logging, another cross-scale dynamic affects seed dispersal. The trees' synchronous mass reproduction requires vast spatial areas. With reduced forest cover, the trees produce less seed—and less viable seed. Furthermore, as El Niño droughts have intensified, fires in the rainforest have worsened. During the 1997–98 El Niño in Indonesia, about 10 million hectares, an area the size of Costa Rica, went up in flames. Smoke from those fires kills vulnerable seedlings and causes health problems for people and wildlife. And it creates another kind of feedback.

Recent research by Daniel Rosenfeld of the Hebrew University of Jerusalem shows that heavy smoke from forest fires can nearly shut down rainfall in tropical areas. The particulate matter of smoke forms so many tiny condensation nuclei that no single water droplet gets big enough to fall. As a result, moisture in the air is spirited away in smoke.

Thus, cross-scale interactions, from global to regional and local climate, are fundamentally altering the rainforest ecosystem. Cross-scale global eco-nomic forces are interacting with regional and local economies to deprive local people of a source of their livelihoods. And climate change is interact-ing with deforestation to hasten the rainforest's decline, undermine human health, and drive species toward extinction, as destructive change cascades through multiple, interacting panarchies.

CHAPTER 6
NATURAL CONGREGATIONS

S tandard ecological theory describes all sorts of natural phenomena as having simple, evenly spaced distributions. Panarchy suggests that while such patterns exist, they may be more the exception than the rule.

"If you look carefully, you see the imprint of various sets of mutually reinforcing ecological processes at different scales," Holling says. These imprints can be found on spatial patterns on a landscape, on temporal patterns in life cycles and behaviors, and on morphological patterns such as body size.

Holling first discovered such an imprint in the weight distribution of birds that prey on the spruce budworm in fir forests. Some 35 species of birds prey on the insect and help keep its population in check. The birds are clumped into five major body mass groups. Species in the same clump forage at similar scales, but vary in their response to environmental changes. Species in different clumps forage at different scales. Small warblers, for example, forage on insect aggregations on branches, larger ground sparrows forage on aggregations on trees, and still-larger grosbeaks forage throughout forest patches. Hence, as budworm populations jump from smaller to larger levels of the panarchy, more and larger avian predator species come in to forage. The size clumps of the birds apparently evolved in concert with the dynamics of forest growth and of insect pest outbreaks, suggesting that clumps are a highly conservative feature, reflecting slow ecological and evolutionary processes that structure panarchies at all levels of scale.

Not only do other animal communities reveal gappy patterns in body mass, but these patterns also reflect the complexity of the ecosystem. For instance, the more architecturally simple the landscape, the fewer body mass clusters. The inhabitants of simple marine sediments are clumped into just three or four size classes. Boreal forests are more complex, with mammal and bird communities clustering into about eight body mass groups, while birds in tropical forest systems form even more clusters.

Clumped structures are vital to ecosystem resilience. This is because an ecosystem can only survive environmental change if it has a variety of species that are capable of supporting key ecological functions as conditions vary. Loss of a species that is important under some conditions will scarcely affect ecosystem functioning if other species can play the same role. If there are no substitutes, however, the loss of some species can trigger a fundamental change from one ecosystem type to another. For example, after overfishing depleted the numbers of large algae-eating fish in Caribbean coral reefs, sea urchins took their place in the food web. The urchins ate the algae, thus maintaining the balance of the coral reef ecosystem. But when an epidemic decimated the sea urchin population, no large grazers remained to consume the algae. The algae proliferated and overgrew the corals, causing massive die-offs of the coral reef.[12] Within-scale and between-scale diversity thus provides crucial overlapping reinforcement of ecosystem function.

Brock and his colleagues have studied the significance of this diversity for agroecosystems (ecological systems whose species mix is transformed to serve agriculture).[13] Agroecosystems may be especially sensitive to the loss of species precisely because they have already been simplified by the exclusion of competing species and predators. This simplification reduces the system's resilience. "The costs of a reduction in resilience include, for example, the cost of the herbicides, pesticides, fertilizers, irrigation and other inputs needed to maintain output in the simplified system. They include the cost of relief where output fails, relocation where soils or water resources have been irreversibly damaged, rehabilitation where damage is reversible, and insurance against crop damage by pest or disease. If the system loses resilience and flips from one state to another, they include forgone output under the new state," Brock writes.

Natural congregations—clumped structures—are significant for another reason as well: they are associated with significant ecological and evolutionary processes. For example, animals that exist at the edge of body size clumps are more likely to be species that act as the indicators and creators of change—species that are endangered, invasive, nomadic, and migratory.[14]

CHAPTER 7
CASCADING CHANGE

The hierarchy theory in ecology stresses the control exerted by large, slow processes. In this view, slow, broad variables constrain, restrict, and dominate: for instance, geology and climate dictate vegetation in an area. "That's true, and it is part of panarchy," Gunderson says. "But there are times when the fast and the small cascade to overcome those controls. Panarchy is a more symmetrical theory."

In panarchies, transformational change can be generated from below or from above. At the same time, larger, slower levels can act to reinforce and sustain the panarchy. Two critical types of change that occur between levels are "revolt" and "remember".

Revolt

In a revolt, processes scale up beyond the local level. A massive forest fire requires not only dry conditions, but also an accumulation of fuel over large scales, so that when lightning strikes, a local ignition will spread to the crown of a tree, then to a patch of forest, then to a stand of trees, and finally to the forest itself. Each step moves the conflagration to a larger and slower level.

Such was the fire that rampaged through New Mexico's Mesa Verde, a national park set up to preserve archeological treasures. July 2000 was a hot month, with almost no rain. Previous efforts to suppress small fires probably contributed to the buildup of high fuel loads in the park. When lightning struck, it was able to spark a fast-moving fire that burned 1,000 acres the first day, 5,000 the next, and then blew out of control. Flames leapt up to 300 feet high as the fire raged through the tinder-dry forest of pinyon pine and Utah juniper. It consumed mountain shrublands of oak and serviceberry. While these shrublands should quickly resprout and recover in just a few years, the pinyon and juniper forests are likely to be colonized by grasses and non-native plants such as thistle. It may take as long as 300 years before the evergreen woodlands are restored.

As in the case of this forest fire, revolt is not a random event. It relies on a large metastructure that has accumulated capital (such as fuel) that allows the revolt to spread. In a revolt, when one level in a panarchy enters the

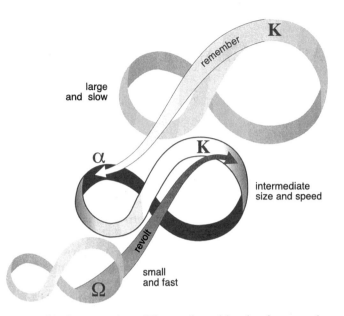

Figure 4. Panarchical connections. Three selected levels of a panarchy are illustrated, to emphasize the two connections that are critical in creating and sustaining adaptive capability. One is the "revolt" connection, which can cause a critical change in one cycle to cascade up to a vulnerable stage in a larger and slower one. The other is the "remember" connection, which facilitates renewal by drawing on the potential that has been accumulated and stored in a larger, slower cycle. An example of the sequence from small and fast, through larger and slower, to largest and slowest for ecosystems can be found in forest-pest dynamics, where small, fast insects affect the slower variable, foliage, which in turn affects the largest and slowest variable, the tree. For institutions, those three speeds might be operational rules, collective choice rules, and constitutional rules (Ostrom 1990; Chapter 5); for economies, individual preferences, markets, and social institutions (Whitaker 1987); for developing nations, markets, infrastructure, and governance (Barro 1997); for societies, allocation mechanisms, norms, and myths (Westley 1995, Chapter 4); for knowledge systems, local knowledge, management practice, and worldview (Gadgil et al. 1993; Berkes 1999; Chapter 5).

phase of creative destruction and collapses, that collapse reverberates upward to transform the next larger and slower level.

Destructive catastrophes also can cascade *down* a panarchy. Chance events external to the adaptive cycle can trigger collapse; extremely large events can overwhelm panarchies despite their sustaining, reinforcing nature.

The asteroid impact thought to have wiped out the dinosaurs and 70% of Earth's species 65 million years ago was such an event. That event, possi-

bly accompanied by massive volcanic eruptions, unraveled the web of interactions within and between panarchical levels over scales from biomes to
species. Some 10 million years of evolutionary change were needed to reestablish the lost diversity.

In human systems, revolutions and state breakdowns can represent such
collapses and transformations of a panarchy. For instance, revolutions can
occur when several segments of society (levels of the panarchy) are in simi-
larly vulnerable stages of the adaptive cycle. Then crisis on one level can
trigger crisis in the others. J. Goldstone describes the English Civil War and
Oliver Cromwell's "Glorious Revolution" in the seventeenth century as the
result of a mounting, reinforcing crises within the governing monarchy, the
elite, and the masses.

Human institutions can crash after periods of success bring about their
own downfall because of the stresses and rigidities that have slowly accumulated. In human as in ecological systems, "A system's increasing dependence
upon the persistence of its existing structure leaves it increasingly vulnerable
to any process or instability that releases its organized capital," says Garry
Peterson, a political ecologist at the University of Wisconsin in Madison.

Organizations and institutions fail to cope with these slow changes
either because they do not recognize them or because no action can be
agreed upon. Disintegration can propagate among several levels of a monolithic culture into systems of competing ideologies. These pave the way for a
new synthesis by visionary or charismatic authority, which in turn can form
an entrenched hierarchy and monolithic culture.[15]

Modern democracies are vulnerable to the same process, but they diffuse
large episodes of creative destruction by creating smaller cycles of renewal
and change through periodic political elections.[16] These smaller cycles are
themselves subject to the adaptive cycle and creative destruction, as seen in
the 2000 presidential race in the United States, when a handful
of contested votes in the State of Florida made it temporarily impossible to
determine the next president. Conflicts and problems in Florida's election led
to a re-evaluation of voting regulations, legislative oversight, and physical
equipment, much of which is now slated for the junk heap and is scheduled to
be replaced by more sophisticated and less potentially subjective technology.

The nested nature of state and federal institutional panarchy, including
the Supreme Court and the U.S. Constitution, largely contained the voting
crisis in Florida, despite its national impact. This shows how the conservative
nature of the panarchy can slow change, while at the same time accumulating
potential and rigidities that can be released—given a large enough disturbance and an entrainment of cycles within the panarchy. "Had the crisis
coincided with other large-scale crisis across the panarchy—widespread un-
employment among the populace, conflict within the military—we could
have had a different outcome," Peterson says.

Cascading change does not have to be catastrophic. Whole panarchies
can be transformed and rejuvenated by novelty that cascades up the levels.

But novelty on this scale is far more than the introduction of a new musical form. It is more akin to the cultural influence of enslaved Africans, who brought with them entire systems of belief that infused new artistic forms and transformed the cultural landscape of the Americas.

Likewise, genetic novelty, the stuff of evolution, is not so much a single mutation of a nucleotide base, but rather multiple combinations generated within genomes by sexual recombination that can open entirely new sets of adaptive paths for natural selection. Today, agricultural geneticists are searching plant genomes for such novelty—complex traits that can confer disease resistance to susceptible crops.

In the case of North America, thousands of years ago, novelty ran wild, transforming entire panarchies and generating new ecosystems. As the ice sheets retreated 15,000 years ago, a protracted reorganization phase of the adaptive cycle commenced and gradually shifted northward.[17] However, ecosystems did not move as integrated wholes. Rather, individual species moved at their own rates and established themselves wherever they could survive based on local conditions. Once established, species that had been strangers to each other set up novel associations. Where chance compatibility existed, key species developed sustaining networks of relationships, laying the basis for new ecosystems. These were consolidated during the phases of rapid growth and conservation that followed and through subsequent adaptive cycles. Today's northern boreal forests, temperate broad-leaf forests, the Everglades of Florida, and the central grasslands all are the result of novelty creating new panarchies and cascading up levels over many iterations of the adaptive cycle.

Panarchical transformations also occur in human systems when novelty cascades up. Two such great creative transformations in human progress were the agricultural revolution 10,000 years ago and the industrial revolution that began around 1750. The latter was not simply triggered by the single invention of the steam engine, but emerged within the context of a whole economy and society that had accumulated a set of rigidities and invented novelties that precipitated, synergized, and directed the transformation.

The same thing is happening with the Internet, as information technology leads a new technological revolution. Its social effects, which may include a significant global impact on industrial organization and the structure of economic activity, employment patterns, culture, and lifestyle, are only beginning to be widely experienced. They may displace some forms of human intellectual activity (in much the same way that the industrial revolution displaced some forms of human physical activity) while enabling others, possibly including far more complex forms of social organization.[18]

Such transformations are rare simply because a unique combination of developments must coalesce: the appearance of novelty on many scales simultaneously and the entrainment of neighboring adaptive cycles within the panarchy.

CHAPTER 8
REMEMBER

Remembrance is the opposite of revolt. Memory and biotic legacies held within higher, slower levels of a panarchy are available to sustain and renew lower, faster levels nested within.

In an ecosystem, biotic memory is at work following a forest fire, when options for renewal draw on the seed bank, physical structures, and surviving species that had accumulated during the forest's slow growth. Mangrove forests in the southern Everglades, subject to flooding, hurricanes, and lightning, provide a unique example. The waterlogged soils there are so electrically conductive that when lightning strikes it kills all the trees within a 10-meter radius. Surrounding forests refurbish the area as seeds enter and new growth begins in the blackened earth. In a common sequel of disturbance, the fierce winds of hurricanes kill tall mangrove trees, stripping their leaves and piling up mud that deprives the roots of oxygen. Regrowth occurs because small areas previously struck by lightning are less affected by the storm's raging winds and escape relatively unscathed. The patches that had been nourished by the surrounding forest then become sources of memory and renewal on a larger scale following the gale.

Institutional memory can operate within human systems as a conservative, sustaining force. Memory is imbedded in libraries, in constitutions, in articles of incorporation. In the Everglades, the use of federal resources for restoration is linked to a more tacit form of memory—the social view of the Everglades as a global icon for conservation. The Everglades has come to be widely viewed as an imperiled wetland whose survival is symbolic of our ability to conserve nature on a planetary scale. Therefore, federal dollars have been made available in huge amounts, Gunderson says.

Institutional memory also resides in individuals. In the case of resource management, the retirement or dismissal of wise and experienced managers may result in an unrecognized loss, while the involvement of such individuals can permit success.

In traditional knowledge systems, respected elders are often the repositories of memory, relied upon to transmit wisdom and practices rooted in generations of experience. For example, Pacific Island societies experience tropical hurricanes about once in a human generation. The only way to transmit the memory of how to respond to such disasters is through the elders. After a hurricane devastated the Pacific island of Tikopia, chiefs directed facility repairs, took measures to reduce opportunities for theft, and redirected labor from fishing to planting. Under the chiefs' leadership, people changed their diets and reduced their ceremonial obligations. They adjusted resource management techniques to include shorter fallow periods between planting cycles and stricter demarcation of land boundaries.[19]

The "wise person" in any successful adaptive management system combines a knowledge of nature's dynamics with a knowledge of the social system. "When you look closely at resource management, it becomes obvious how arbitrary the separation is between people and nature," says Fikret Berkes, a professor of natural resource management at the University of Manitoba.

CHAPTER 9
THE ADAPTIVE CYCLE AND LOCAL KNOWLEDGE

P anarchy asks how novelty occurs, is encouraged, or is suppressed. The same question can be asked of wisdom. How does wisdom accumulate? How is it able to manifest itself within natural resource management systems? How is it transferred and deepened?

Berkes and Carl Folke, an ecologist at Stockholm University, examine this question in the context of traditional knowledge systems. They have found traditional knowledge systems that challenge the notion that there is only one overt world of resource management: the rationalist, Newtonian-clockwork, conventional world of government managers.[20]

"In traditional systems, knowledge is carried, transmitted, and renewed in ways that increase the resilience of the adaptive cycle," Berkes says. "Indigenous systems could provide insights for conventional management systems, and the two could complement each other in many management efforts."

Indigenous peoples who live off wildlife, fish, and forests create their own knowledge from their observations and their understanding of ecology, based on generations of trial-and-error experience. Local and indigenous practices are based on linkages and feedbacks between natural systems and social systems rather than upon their arbitrary separation. Many practices are attuned to the four phases of the adaptive cycle. Some concern the cycle's front loop, the phases of rapid growth and conservation, while others correspond to the back loop of creative destruction and reorganization.

The front loop of the cycle may be thought of as a succession sequence, from the initial few pioneers—the first grasses that appear after a forest fire during the rapid growth phase—to the mature and complex community, such as a climax forest, in the conservation phase. In classical succession theory, the climax is the state the system would attain and maintain if not disturbed, rather than a transitional phase in a continuous cycle. Conventional resource management has largely concerned itself with front loop processes, seeking to determine the equilibrium level of a population in its environment, and from that estimating the surplus available for harvest. Therefore, it focuses on quantitative population models for management decision-making.

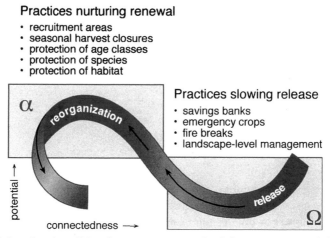

Practices nurturing renewal
- recruitment areas
- seasonal harvest closures
- protection of age classes
- protection of species
- protection of habitat

Practices slowing release
- savings banks
- emergency crops
- fire breaks
- landscape-level management

Figure 5. Local and traditional management or "back-loop" practices of the release and reorganization phases of the adaptive renewal cycle. Some practices work to slow down the rate of release, while others nurture sources of renewal.

In the eastern subarctic region of Canada, the Cree Indians manage caribou by monitoring much the same information as does Western science —geographic distributions, sex and age composition of the herd, fat deposits in caribou, and the like. However, the Cree, like other northern peoples from Labrador to Alaska, neither produce nor use estimates of population size. Rather, their system provides hunters with a qualitative mental model that indicates the population trend—placing special emphasis on an assessment of caribou fat.

Monitoring fat content makes sense because it provides an index of health of both the individual animal and the herd as a whole. It integrates the combined effects of a number of environmental factors, such as predation and the condition of the feeding range. This qualitative approach reveals whether the caribou herd is increasing or decreasing in size, and gaining or losing health and fitness, without requiring an actual count of the population to make management decisions.

Other traditional knowledge systems also rely on qualitative indicators. For instance, indigenous people in the Pacific Northwest managed salmon harvests through a qualitative assessment of the intensity of the spring salmon run. Based on their observations and stored knowledge from previous years, religious leaders would assess the run and use rituals to establish taboo fishing periods of varying durations.

During the back loop of an ecosystem's adaptive cycle, disturbance typically exceeds resilience, and triggers rapid transformation. Thus, periods of gradual change and rapid transformation complement one another. Many

local and traditional practices create disturbance. By mimicking natural per-
turbations they avoid the kind of catastrophic disturbance that can jump
scales and restructure the panarchy. As an example, Australian aborigines,
Californian Amerindians, and Canadian Amerindians used fire management
to open up clearings, corridors, and windfall areas. These practices improved
habitat for grazing wildlife and waterfowl, and consequently improved
hunting.

In contrast, conventional resource management practices tend to avoid
rapid transformation, instead supporting the front loop phases of gradual
change. This kind of management restricts ecosystem diversity and variability and diminishes resilience. However, resource managers who operate
based on Western science are beginning to adopt traditional back loop
processes, such as controlled burns of forested areas.

Traditional practices also may reduce the effects of natural disturbance
and contribute to ecosystem recovery during the back loop of the cycle. One
example is the designation and protection of sacred groves in India. These
groves, established for religious purposes, serve as a gene bank and seed bank
after storms and when other forests have been felled.[21] Sacred areas used to
be common in terrestrial ecosystems from the Americas to Africa. Sacred
marine areas, although less common, can be found in the South Pacific and
existed along the East African coast until the 1950s.

Madhav Gadgil, an ecologist at the Indian Institute of Science in
Bangalore, India, describes five "rules of thumb" in indigenous systems that
parallel scientific systems of conservation: total protection of certain
species; protection of vulnerable life history stages; protection of specific
habitats; temporal restriction of the harvest; and monitoring ecosystem
change. Each of these practices nurtures sources of renewal and thereby
aids ecosystem reorganization and recovery.

CHAPTER 10
HOW DO HUMAN AND NATURAL SYSTEMS DIFFER?

"The elephant is good for thinking," says one African proverb. So, too, is panarchy. It is a tool for thought in creating sustainable futures. To further hone this tool, it is worth considering not only the similarities, but also the differences between human and ecological system panarchies.

The most fundamental difference is the way in which inventions are accumulated and transferred over time—through genes in organisms, through mutually reinforcing patterns in ecosystems, and through communication in human systems.[22]

Like genetic inheritance, ecosystem behavior derives from the past. Its dynamics are the result of past evolutionary successes and of the complex, mutually reinforcing relationships between species, their environment, and the environments they create. Memory is in the form of biotic legacies. While ecological structures can be vast and complex, such as a coral reef, they cannot willfully create alternative structures and processes to cope with environmental changes imposed by human activities.

But human cognition and communication permit a higher level of self-organization than that found in ecosystems. Organizations can adjust their functioning, from rigid to fluid, as the need arises. Humans can change their roles depending on the circumstances. While nature's capacity for remembrance resides in the form of biotic legacies, humans and human systems have the capacity of consciousness, communication, and reflection. Furthermore, human systems have the capacity to design institutions (such as futures markets) and economic models that anticipate the future and use what is learned to adjust actions in the present.

The Dimension of Human Cognition

Space and time are two dimensions that shape ecological and human systems alike. But the power of cognition adds a third dimension that greatly helps to structure human social system dynamics. The ability to construct and manipulate symbols, most obviously words, creates "structures of signification"—a hierarchy of abstraction.

In *Panarchy*, Francis Westley at McGill University in Montreal and his colleagues analyze the role of this third dimension. Driven to make sense of the world, humans use communication, language, and symbols to invent and reinvent meaningful order and then act in accordance with that invented world as if it were real.

Systems of meaning provide a "virtual reality" that can enhance resilience, providing cohesion during times of material shortage or social crisis—when things fall apart. As long as the structures of signification stay in place the whole system will not transform radically, but rather will return to a previous equilibrium. As Westley notes, the opposite is also true: if meaning is lost human systems seem unable to recover.

Human societies manifest their virtual realities in technology and the construction of technical systems, which in turn can come to appear to be outside of human control. And they do this on an Earth-changing scale, since the human ability to exploit a variety of niches and scales far surpasses that of other species.

The capacity for representation, for communication, and for making meaning seems to drive both the processes of maintaining system integrity and of dealing with change. Yet this human ability has limits when applied to complex problems of the environment.

Failures of Foresight

Despite the human capacity for foresight, people have difficulty handling variables that change slowly, such as rising ocean levels, the decay of infrastructure, or the depletion of an aquifer. As Brian Walker and Nick Abel point out in *Panarchy*, we are much better at adapting to changes in fast variables such as grass growth, animal numbers, and stock prices, or even medium-term processes such as the spread of insidious diseases like AIDS. Failure to deal with time horizons of years is a common cause of human failures in decision-making.[23]

The ability to think and plan ahead is part of the human brain's "executive function." Executive function is the result of the maturation of a complex adaptive system—the human brain—and of the prefrontal cortex in particular, which is evolutionarily young relative to other brain structures such as those that make up the limbic system. Therefore, inconsistencies or limitations in executive function are perhaps in part a reflection of our own stage of evolution. Human problem solving seems better attuned to processes that do not exceed the length of a human generation and the duration of human institutional or cultural memory.

In addition, humans tend to solve problems one time scale at a time. Juggling a monthly budget or saving for a child's college education and one's own future retirement is significantly more complex than balancing a checkbook. While the focus on one time-scale may lead to short-term success in a narrow domain, it often limits longer-term options and resilience, and creates spin-off problems that appear to be a long way off.

The ability to create novel futures through technology is itself a double-edged sword. Humans often fail to build adaptive capacities into their technologies, tending instead to create technologies that address a single scale or variable and disregard its impact on other parts of the system. This creates new problems farther down the line.

For example, the industrial production of nitrogen-rich fertilizers allowed farmers to increase crop yield on a massive scale beginning with the Green Revolution. Today, human activities (including the burning of fossil fuels) contribute more to the global supply of fixed nitrogen than do natural processes.[24] Although ecosystems can absorb a limited amount of additional nitrogen by producing more plant mass, above a certain threshold, excess nitrogen can overwhelm the natural nitrogen cycle. Unexpected side effects of increased nitrogen use have included toxic levels of nitrate in groundwater, widespread acidification of ecosystems, loss of other soil nutrients such as calcium and magnesium, changes in the number and kind of species in affected ecosystems, and the production of toxic algae blooms in coastal waters.

In addition, foresight can be narrowly focused on private gain—protecting one's position rather than furthering social goals. In management agencies, where entrenched interests can manipulate information for narrow purposes, it is rare for novelty to cascade up through a panarchy to change a policy. In such cases of tight bureaucratic control, panarchical change only occurs when a triggering event unlocks the social and political gridlock of larger scales.

CHAPTER 11
CHALLENGES OF ADAPTIVE MANAGEMENT

In managing ecosystems and natural resources, restorative changes that can cascade through a panarchy come as a conscious act of wise, purposive design and implementation. Such events are rare, as even when the ecological components of a system are thoroughly understood, chance events and changing human dynamics can lead to surprises. The management history of Lake Mendota in Madison, Wisconsin, is a case in point.

If you have never been to Lake Mendota, you can see it on the Internet (www.soils.wisc.edu/asig/webcam.html). A live web cam set up by soil scientists at the University of Wisconsin provides an ever-changing panorama of the lake. Just beyond dormitory rooftops, clouds billow by, the sun rises and sets, and sailboats cross. Occasionally, the limnology department's research trawler *Limnos* motors by.

Nearly 10,000 acres wide and up to 83 feet deep, Lake Mendota has been called the most studied lake in the world. Its changing waters have offered insights into how the adaptive cycle works in lakes, and how, through eutrophication, clear, healthy lakes can become murky. Efforts to maintain the quality of the lake, which is a favorite recreation spot in the region, provide insights into the challenges of ecosystem management.

Clear lakes can shift into a turbid, algae-covered state due to an overdose of nutrients such as phosphorus. Animal waste, agricultural and lawn fertilizers, sewage, and soil erosion all can contribute to excess phosphorus. In 1995, analysts calculated that Lake Mendota receives 75,000 pounds of phosphorus a year. Such nutrients stimulate the growth of phytoplankton, which trigger dramatic changes in the underwater world. Algae blocks light, killing submerged plants that normally grow in clear water. Without the plants, countless small animals that lived on and among their swaying stems die. Next to go are fish that depend on the small animals. Finally, migrating birds fail to visit the lake, and bird numbers typically drop by one or two orders of magnitude when a lake becomes turbid.[25]

The devastating effect of overloading lakes with waste nutrients is a major problem throughout agricultural and urban areas. Although deeper lakes may recover more easily given good management, shallow lakes rarely improve. Even when the nutrient load is reduced to well below the level that caused the collapse, shallow lakes tend to remain turbid and eutrophic.

Lake Mendota sometimes turns the color of pea soup as a scum of blue-green algae covers its surface, but it has not yet become permanently degraded wastewater. Scientists and government agencies have worked long and hard to prevent the lake from crossing a threshold of nutrient loading beyond which water quality would be permanently damaged. Limnologist Steve Carpenter at the University of Wisconsin notes that Lake Mendota is extremely fragile, and that a $10 million mistake today could easily grow into a $100 million problem if left uncorrected. "I see this lake as a symbol of our ability to manage the other 15,000 lakes in this state," he says.[26]

Carpenter and his Resilience Alliance colleagues have analyzed how the adaptive cycle in the ecosystem and management has shaped the lake's recent history. They write that Lake Mendota illustrates a long, slow slide into the turbid state, followed by repeated attempts to restore the clear-water state, in a sequence of four adaptive cycles. With each cycle, however, management has learned important lessons and has sought to apply them to the changing circumstances.

The lake's watershed is a 230-square-mile drainage basin that includes most of the city of Madison, as well as numerous other cities and towns. Almost 60% of the watershed is agricultural; 20% is urban, and about 6% is wetlands, with half of the area's wetlands having been lost in the last century. The adaptive cycles of the lake and of management have both reflected and influenced changes in surrounding land use. Carpenter summarizes the five linked management-ecosystem adaptive cycles in Lake Mendota's recent history as follows:[27]

Cycle 1. This was initiated when European immigrants began to settle in the watershed around 1840. For the first time, rich prairie soils were ploughed, sending soil and runoff into the lake, and bringing about an immediate loss of resilience. The initial rapid growth in population and farming (the r phase of the adaptive cycle) was followed by a long transition through the conservation phase. Decades passed during which there was little increase in farming, but steady, slow growth of the human population. This led to increased sewage effluent and a growing nutrient load in the lake. Then, shortly after World War II, agriculture and urbanization intensified sharply. This triggered a collapse in water quality and the phase of creative destruction. The following reorganization phase lasted through a decade of public debate.

Cycle 2. Concluding the debate, a plan to divert sewage effluents from the lake launched the second cycle of lake management. Diversion was completed in 1971, but increases in water quality were minor. The phosphorus (P) from sewage effluent had been replaced by nonpoint P inputs due to increased fertilizer use, increased dairy herd sizes, increased P content of soils, and sprawling urbanization. The water quality of the lake was declining, evidenced by another phase of creative destruction, when algae blooms again covered the lake. By 1978 it was clear that although sewage diversion was a step forward, it had not achieved lake managers' water quality goals. The recognition of this fact led to an institutional transition that ended the second management cycle.

Cycle 3. Lake managers quickly tackled the nonpoint pollution problem, initiating Cycle 3. They designed a project to decrease nonpoint P loss from the steepest and most erosion-prone subwatershed of Lake Mendota. But the project failed to motivate farmer participation, limiting its success. Cycle 3 ended quickly. Again, the critical events were not ecological, but rather an institutional recognition that another approach was needed.

Cycle 4. The reorganization phase that initiated Cycle 4 came in the 1980s when managers realized that stocking the lake with large numbers of walleye might improve water quality and eliminate the need for costly and controversial interventions on private land. Previous experiments in northern Wisconsin lakes had shown that the introduction of walleyes could alter the food web in a way that would reduce concentrations of algae, while maintaining the system's integrity.

The project scored some initial successes, and water quality improved. But it also attracted the interest of anglers eager to catch walleye, and the numbers of people fishing on the lake jumped seven-fold. Despite a restrictive catch limit, the unexpected growth of fishing pressure drove down the number of walleyes, reducing their effectiveness in controlling algae blooms.

The final blow came with heavy summer rains that caused massive erosion from construction sites into the lake. Phosphorus inputs spiked to the largest ever measured as the rains of 1993 triggered the creative destruction phase that brought an end to both the ecological and management cycles. The dramatic turn for the worse made it clear that the main problem confronting the lake was nonpoint pollution with origins dispersed across the landscape.

Cycle 5. The fifth management cycle began in 1998 with a plan to cut nonpoint pollution in half. Intensive research by both agency and university scientists culminated in an aggressive plan called the Priority Lake Project. It began with an assessment of pollutant loads from all the main nonpoint sources in the watershed, including barnyards, agricultural lands, and urban areas. For instance, an inventory of 402 barnyard and animal lots estimated that they generate 20,064 pounds of phosphorus annually, with 75% ending up in Lake Mendota. Under this new program, implementation of nonpoint pollution controls will continue through at least 2006. Limnologists expect that impacts on water quality should be evident by 2010, although it will take more than 20 years to fully evaluate the results.

While the 10-year Priority Lake Project focuses on reducing nonpoint pollution, Carpenter notes another problem: urban sprawl is increasing in the watershed, covering still more land with cement and asphalt, which are impervious to rainfall. Runoff threatens to increase the intensity and frequency of floods. At the same time, soil P levels in the watershed are stabilizing or declining, increasing the resilience of the lake's clear-water state. The improvement may be the result of farmers switching to tillage practices that reduce erosion. Thus, even as progress is made on one front, the underlying causes of Lake Mendota's problem may be shifting from intensive agriculture to urban sprawl and erosion from construction sites.

Adaptive management "is a question of creating the right links, at the right time around the right issues to create a responsive system," Holling writes in *Panarchy*.[28] It is not simply a matter of identifying the best practices or institutional arrangements, and even less so a matter of controlling a single variable with the goal of maintaining the ecosystem in a steady state. In dynamic systems, only uncertainty is certain, and it arises from multiple causes.

As in the last several cycles of Lake Mendota, an adaptive management approach applies an understanding of the adaptive cycle to identify the point at which interventions may be able to shift the cycle into a more sustainable mode. It does so, in part, by designing actions that accept uncertainty while furthering learning. Modeling techniques can help sort through multiple hypotheses, each of which have their own social and management implications. Management experiments are then designed to test the most likely hypotheses in the real world, and to learn from their outcomes.

A foundation of adaptive management is a planned sequence of experimental treatments. "You can't just go in, change things, and then evaluate it.

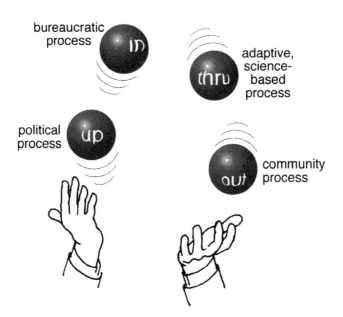

Figure 6. Four processes represented as four balls, which an effective manager who seeks to harness complexity must juggle simultaneously.

Adaptive management is not about 'try it and see,' but about carefully planned experimental evaluation of policies," says Resilience Alliance member Carl Walters, an ecologist at the University of British Columbia.

Such experiments can be risky economically and politically. For instance, when Mendota mangers stocked the lake with walleye, they knew they were taking a risk. The experiment was expensive, untested on such a large scale, and politically charged because of the many constituencies involved. Such experiments require the active involvement and education of stakeholders, not only to support the experiment, but, when possible, to participate in it. Although the Lake Mendota experiment began with the involvement of the angler community, it proved difficult to devote the human management resources needed to sustain that involvement, Carpenter says.

In *Panarchy*, Westley describes such an experiment on another Wisconsin lake. Resource managers worked with salmon fishermen in an aquaculture experiment to grow caged salmon. Although managers doubted the experiment would work, they agreed to carry it out on the insistence of the anglers, who were passionately involved in the effort. After two years of failure, the fishermen themselves decided to end the experiment. During its course, anglers and resource managers built strong ties, mutual respect and a deeper understanding of the resources they sought to manage.[29]

In the case of such adaptive management, "the resilience of the ecological system provides 'insurance' within which managers can affordably fail and learn while applying policies and practices. The social equivalent of ecological resilience, human adaptive capacity resides in the ability to confront uncertainty and develop understanding of what contributes to loss of ecological resilience," writes Holling.[30]

CHAPTER 12
PANARCHY AND THE ECONOMICS OF NATURAL RESOURCES

Today, the Lake Mendota project includes substantial cost-sharing incentives for farmer participation, funds to enforce regulations for control of erosion from construction sites, and funds to purchase riparian easements and wetlands for restoration. Missing, however, are economic mechanisms that take into account the cost of a loss of resilience. For instance, the cost of maintaining water quality is not figured into the price of water. Nor is there a system of incentives that encourages farmers and others to cut back on their use of nutrients. Carpenter suggests that one way to do this would be to create a market for rights to emit phosphorus.

Such a system, says economist Brock, would address a key aspect of linked economic and ecological systems: the design of institutions that force people to pay the full social cost of their actions. In such a "full-costing" system, those who create the costs will bear the costs. When a farmer decides to fertilize her fields and the fertilizer washes into a lake, or when an industry emits large amounts of carbon dioxide into the atmosphere, each is off-loading part of the cost of doing business onto society at large. The problem has been described as "CCPP"—Commonize the Cost, Privatize the Profit. To meet the challenge of full-costing, new mechanisms are needed that go beyond reliance on market prices.

In *Panarchy*, Brock and others discuss the types of mechanisms that could work, given the tight linkages between the economy and ecosystems that are capable of unpredictable collapse.[31] "This dynamic linkage creates a new focus in economics," Brock says. Faced with potential collapse, standard assumptions about the ability of the market to function based solely on prices and the principles of supply and demand can fail.

For instance, farmers regulate their use of a pesticide in part based on its price. If the price is low enough, farmers may use so much pesticide that the excess runoff into a nearby lake induces a collapse of the ecosystem. After the collapse, merely increasing the price of the pesticide will not induce the lake to flip back to its previous, healthy state. "The price would have to be put way up and held up for a long time before you would get a payoff in terms of cleaning the lake," explains economist Charles Perring at the University of York in England. In some cases, no price would ever be high enough to bring the lake back.

Instead of relying solely on prices in markets that only reflect private costs, Brock advocates the use of flexible market-based mechanisms in which market prices are adjusted to reflect social costs. Taxes would be one such mechanism. For example, Brock points to the problem of polluting emissions and crash damage to smaller vehicles caused by sports utility vehicles, which are much heavier and consume more gasoline than a fuel-efficient car like a Honda hybrid. A formula could be used to calculate the social damage done by driving 100,000 miles in an SUV versus a smaller, more fuel-efficient car, and a corresponding social tax could be charged on the purchase of the SUV. Brock explains: "When a person buys a large SUV, he'll have to pay, perhaps, $2,000 more. If he has a lot of kids that he takes camping, a large SUV may be important to him and he'll pay the extra money. On the other had, if large SUVs are priced out of the market, that is as it should be because the corrected price would reflect the true social costs." Such a tax would be a flexible mechanism that would allow people to make their own decisions about the type of car they want to drive while also discouraging polluting by requiring that people cover the full costs of their transportation choice.

The potential for long-lasting or even irreversible change in ecosystems also suggests the need for increased regulatory use of the precautionary principle. This principle suggests that when an activity, technology, or product poses risks to human health or the environment, precautionary measures should be taken even if some cause-and-effect relationships have not been fully established scientifically. In this context, the proponent of an activity with a potentially high social cost, rather than the public, should bear more of the burden of proof than is usual in many cases. In addition, new technologies should be forced to cover the full social costs of their use.

"In the U.S. it seems that the burden of proof is on the people who want to regulate activities that are obviously off-loading costs onto others," Brock says. For example, consider the debate over the effects of genetically modified "Bt" corn. It has been engineered to contain genetic material derived from a soil bacterium, *Bacillus thuringiensis* (Bt), which produces a protein that is toxic to some insects, including the European corn borer. Widespread use of Bt corn by conventional agriculture may lead to increased resistance to the toxin in the insect pests. To combat resistance, farmers would have to use more and different pesticides. This would impose an uncompensated cost on organic farmers, who cannot use pesticides. This is an example of an uncompensated, spillover cost onto organic producers imposed by the producers of genetically modified foods.

In cases like this, proper use of the precautionary principle would reverse at least part of the burden of proof; producers of genetically modified crops would have to show that they are not causing harm, rather than organic farmers showing that they are. An even greater burden of proof should be placed on activities that may lead to a loss of ecosystem resilience—especially in cases where there is increased probability of collapses that are hard to reverse. To put it another way, the burden of proof should reflect the losses to society and the higher costs associated with that potential ecosystem collapse.

CHAPTER 13
LEARNING: AN END AND A BEGINNING

Three types of change take place in panarchies, and each generates its own kind of learning: incremental, lurching, and transformational.[32] Incremental change and learning occurs during the early, predictable stages of the adaptive cycle, as a system moves through rapid growth to conservation. During these phases, managers assume that their models are correct. They collect data and information to update these models. In bureaucratically dominated resource systems, self-referential professionals and technocrats typically view dealing with this type of change and learning as problem solution.[33]

Abrupt change and spasmodic learning occurs during the transitions between the conservation, creative destruction, and renewal phases of the adaptive cycle. Now learning is more immediate and charged. It can take place in the midst of an environmental crisis, where policy failure is undeniable. This type of change and learning is episodic, discontinuous, and surprising. It reveals the inadequacies of the underlying model, which is subsequently questioned and rejected. The problem is reformulated. In bureaucratic resource systems, outside groups or charismatic individuals who can integrate knowledge and interest groups can facilitate this type of learning.

Transformational change is the most dramatic and requires the deepest type of learning. It takes place as cascades of change reconfigure panarchies. The surprise is caused by cross-scale interactions or suites of novelty that ricochet through the system as it reorganizes around alternate sets of mutually reinforcing processes. In these cases, learning involves solving problems of great complexity in a situation of great uncertainty. It involves several levels in a panarchy, not simply one. Not only are new models developed, but also new paradigms.

In *Panarchy*, Holling and Gunderson both predict and present such a new paradigm. Approaches that emphasize a narrow scale of action that seeks an unattainable stability to achieve fragmented goals shift towards approaches that emphasize cross-scale interactions and living with true uncertainty and surprise. In making this shift, "The emphasis should be on flexible institutions and human organizations that can build adaptive capacity in synergy with ecosystem dynamics," Holling and Gunderson write.

We live in a time when gales of change are transforming ecosystems and social systems. Viewed within the framework of panarchy, it is a time when the window for constructive change is open at several scales. Ours is a time of opportunity and of peril. Perhaps the greatest peril is to let the opportunity pass.

NOTES

1. Curran, L.M., Caniago, I., Paoli, G.D., Astianti, D., Kusneti, M., Leighton, M., Nirarita, C.E., & Haeruman, H. 1999. Impact of El Niño and Logging on Canopy Tree Recruitment in Borneo. *Science* 286 (5447), p. 2184. And Wuethrich, B. 2000. Combined insults spell trouble for rainforests. *Science* 289 (5476). 35–37.

2. Jackson, J. B. C. 2001. What was natural in the coastal oceans? *Proceedings of the National Academy of Sciences.* 98, 5411–5418.

3. McNeely, J.A., & Scherr, S.J. (Forthcoming). *Common Ground, Common Future: Ecoagriculture Strategies to Help Feed the World and Save Wild Biodiversity.* IUCN and Future Harvest.

4. Holling, C.S., & Gunderson, L.H. 2001. Resilience and adaptive cycles. In L.H. Gunderson & C.S. Holling (Editors), *Panarchy: Understanding Transformations in Human and Natural Systems.* Washington, D.C.: Island Press.

5. Gunderson, L.H., Holling, C.S., & Peterson, G.D. 2001. Surprises and sustainability: Cycles of renewal in the Everglades. In L.H. Gunderson & C.S. Holling, *Panarchy.*

6. Blake, J.C. 1998. *Red Grass River: A Legend.* New York: Avon Books.

7. Dugger, A. The South Florida Everglades Restoration Project. *http://www.ce.utexas.edu/prof/maidment/grad/dugger/GLADES/glades. html.*

8. Ibid.

9. Gunderson, L.H., et al. Surprises.

10. Ibid.

11. Klausmeier, C.A. 1999. Regular and irregular patterns in semiarid vegetation. *Science* 284, 1826–1828.

12. Jackson, J. B. C. What was natural in the coastal oceans?

13. Brock, W.A., Mäler, K.G., & Perrings, C. 2001. Resilience and sustainability: The economic analysis of nonlinear dynamic systems. In L.H. Gunderson & C.S. Holling, *Panarchy.*

14. Peterson, G. Ten Conclusions from the Resilience Project. *http://www.sustainablefutures.net/rNetFindings.html.*

15. Holling, C.S., Gunderson, L.H., & Peterson, G.D. 2001. Sustainability and panarchies. In L.H. Gunderson & C.S. Holling, *Panarchy*.

16. Ibid.

17. Ibid.

18. Gallopín, G.C. 2001. Planning for resilience: Scenarios, surprises, and branch points. In L.H. Gunderson & C.S. Holling, *Panarchy*.

19. Berkes, F., & Folke, C. 2001. Back to the future: Ecosystem dynamics and local knowledge. In L.H. Gunderson & C.S. Holling, *Panarchy*.

20. Ibid.

21. Gadgil, M. 2001. Sacred groves: Securing a recruitment of seeds and maintaining landscape patchiness. Quoted in L.H. Gunderson & C.S. Holling, *Panarchy*, Box 5.1.

22. Holling, C.S., et al. Sustainability and panarchies.

23. Walker, B. & Abel, N. 2001. Resilient rangelands—Adaptation in complex systems. In L.H. Gunderson & C.S. Holling, *Panarchy*.

24. World Resources Institute. 1998–99. Nutrient Overload: Unbalancing the Global Nitrogen Cycle. *http://earthtrends.wri.org/ conditions_trends/ feature_select_action.cfm?theme=2*

25. Scheffer, M., Westley, F., Brock, W.A., & Holmgren, M. 2001. Dynamic interaction of societies and ecosystems—Linking theories from ecology, economy, and sociology. In L.H. Gunderson & C.S. Holling, *Panarchy*.

26. Mattmiller, B. 1999. Lake Mendota Teems with Teaching and Research Efforts—and Algae. *News@UW-Madison*. Posted 10/05/99. http://www.news.wisc.edu/view.html?get=1990.

27. Carpenter, S., Walker, B., Anderies, J.M., and Abel, N. 2001. From metaphor to measurement: Resilience of what to what. *Ecosystems:* in press.

28. Holling, C.S., Carpenter, S.R., Brock, W.A., & Gunderson, L.H. 2001. Discoveries for sustainable futures. In L.H. Gunderson & C.S. Holling, *Panarchy*.

29. Westley, F. 2001. The devil in the dynamics. In L.H. Gunderson & C.S. Holling, *Panarchy*.

30. Holling, C.S., et al. Discoveries.

31. Brock, W.A., et al. 2001. Resilience and sustainability. In L.H. Gunderson & C.S. Holling, *Panarchy*.

32. Holling, C.S., et al. Discoveries.

33. Westley, F. The devil.